JUST ADD WATER

EASY TECHNIQUES & EVERYDAY IDEAS FOR INSPIRING FLOWER ARRANGEMENTS

CYNTHIA GAYLIN BIGONY

PHOTOGRAPHS BY VIVIAN JOHNSON

WEST
MARGIN
PRESS

Library of Congress Cataloging-in-Publication Data is on file

ISBN: 9781513262888 (hardbound) | 9781513262895 (e-book)

Indexed by Sam Arnold-Boyd

Proudly distributed by Ingram Publisher Services

Printed in China
1 2 3 4 5

Published by West Margin Press

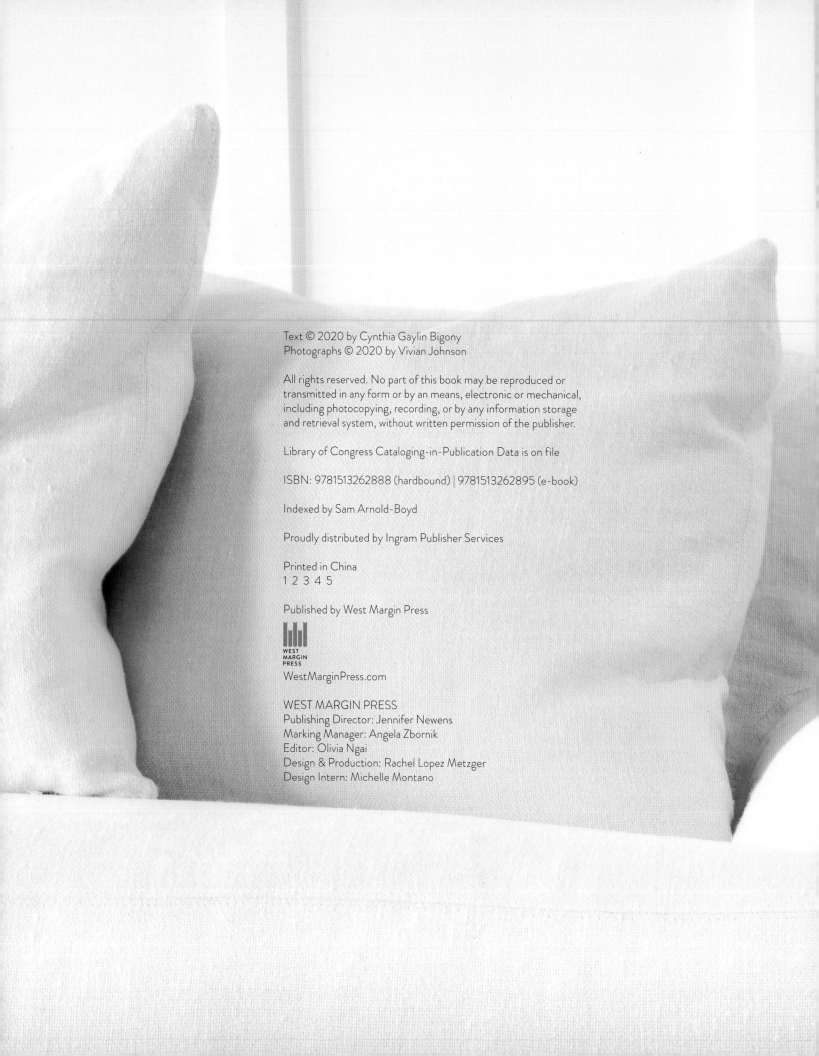

WEST
MARGIN
PRESS
WestMarginPress.com

WEST MARGIN PRESS
Publishing Director: Jennifer Newens
Marking Manager: Angela Zbornik
Editor: Olivia Ngai
Design & Production: Rachel Lopez Metzger
Design Intern: Michelle Montano

To Randy, Kendell, Courtney, Taylor, and pups Charlie & Lola
You are my everything.

A walk on the beach, a good book before bed, or fresh flowers in my home. A great life isn't about great huge things; it's about small things that make a big difference.

Working with flowers is very nurturing. It transports me to a zone of joy I don't want to leave.

I remember my first "aha" moment. It was a late afternoon and I had just finished unloading groceries. It had been a long day and I was tired. Looking over at the flowers I just purchased still made me smile, so I gathered my energy, picked up the blooms, and began prepping. I removed the leaves, carefully recut the stems, and placed each flower into a vase in a way that looked pretty to me.

How was this different from any other time? I became aware of the profound calm I felt as I worked with my hands. As I claimed this moment for myself, the noise of three young children playing with our dogs grew faint. I had found peaceful minutes of restorative joy, an active meditation that energized me!

I am grateful for these simple moments and appreciate them often. However, I also know that when it comes to real life and my hectic schedule, fast and easy flowers are a must. Of course, I love sitting down with an elaborate book showcasing over-the-top party ideas. However, this is not that book.

This is a book with beautiful yet simple designs that anyone can recreate in a pinch. It encourages you to set aside personal time to recharge your soul by working with your hands. Touching nature to nurture.

So many people I meet say, "I'm terrible at arranging flowers." This simply is not true! I wrote this book to demystify the process of creating professional-looking arrangements with ease. If you can hold a pair of scissors, I promise you can create beautiful flower arrangements. Keep in mind that when it comes to creativity, the number one rule is there are no rules. So with simple techniques to guide you and easy ideas to try, you can put your own spin on these, or just copy what you see and enjoy.

My goal for you is to open this book with *curiosity* and close it by saying, *"I can do that!"* Because it's the small things in life that really matter, and sometimes all you need to do is just add water!

XO

Cynthia

GETTING STARTED

ESSENTIAL
tools & materials

This book is about simple designs for everyday living—and that's for your tools and materials too! You don't need anything fancy, just some easy items found in your home for quick access. There are also some handy optional materials that you can find online or at your local craft store.

SCISSORS A pair of sharp flower scissors is ideal, but I will admit, I usually grab kitchen scissors. This is a must for cutting stems and removing lower leaves to keep the water clean.

RUBBER BANDS I like saving the thick rubber bands that come with the asparagus from the grocery store as well as buying different sizes of clear hair bands. A variety of bands will be useful to have.

TAPE Tape is my secret weapon when creating invisible grids to keep flowers and stems in place. Waterproof tape is nice, but regular tape works fine.

RAFFIA I always have raffia in my gift-wrapping drawer. I love the organic, simple effect, which works perfect for flower design. I use it in arrangements to hide rubber bands or to secure leaning orchids.

VASES
& containers

I used to have a pantry full of vases from flowers I had received as gifts, or that I bought just because they were pretty. Here's what I learned: keep only what you love.

Each room in your home has one focal point. I tend to grab the same one or two vases for these spots because I know the scale and lighting are perfect for them. When you buy a new vase, make sure you know where it will go and then edit what you have. Less is more, which makes life less complicated.

On that note, flower vases are hidden all throughout your home. Trust me, I have so much fun thinking outside the box and what's already in my house that can be repurposed for flower containers. It's fun, creative, and unexpected!

So where can you find them? The kitchen is a good place to start. How about a copper meatloaf pan or white casserole dish? A white water pitcher, a glass wine carafe, or a silver ice bucket? Start saving your glass creamer bottles so that when you have enough, you can parade them down the dining table or make a circle with them and fill with garden flowers. Candy and nut dishes are great for floating flowers. Seeing the roots from several small herb plants placed inside drinking glasses work perfect for an inside garden.

Look around your living room and notice the candle holders. These come in many sizes and colors and always look pretty for small bouquets. My go-to glass hurricane candle holder looks beautiful for floating flowers. And if I like a pretty retail bag, why not reuse it for a flowering plant?

Don't forget the bathroom. I've used a decorative trash can and a toothbrush holders as flower vases because they're the perfect size for a sweet bouquet on the sink.

You can find vases anywhere. If it can hold flowers, that's all you really need.

CHOOSING
the container

Cute bags, boxes,
baskets, trash bins

Milk bottles, drinking glasses,
yogurt cups

Here are some fun containers hidden around my house that I like to use for flowers. Repurposing home items provides endless options to display flowers in unexpected ways. It's convenient and easy. Give it a try!

Glass bowls and hurricane
candle holders

Toothbrush holders
or any small container

Water pitchers,
carafes, ice buckets

FILLING
the container

A bag is perfect for a potted plant. If the bag is too deep, put something inside to raise up the flowers so they just peek over top.

These milk bottles displayed together in a circle are a clever way to bring your garden into your home.

Once you find a container you like, it can inform the type of flowers you put in it, or vice versa. Small containers work well for simple posies, and big containers are great for large arrangements. But just go with what inspires you.

Floating flowers are simple and beautiful in clear containers. Play with the waterline to create a variety of looks for your flowers.

Small containers like this toothbrush holder work beautifully together with posy bouquets.

These are lovely statement vessels. Let flowers like these tulips and jasmine hang over the container for movement.

TIPS & TRICKS

There are so many things I wish had known from the beginning about flowers. Let me save you some time! Here are some top tips for getting started.

Use a baby bottle brush to clean hard-to-reach, narrow flower vases.

To clean watermarks in those narrow, hard-to-clean vases, roll up a sheet of paper towel and slip it in the opening. The paper towel absorbs the water and leaves the container free from water spots.

Notice how the word "hydrangea" sounds like hydrate? These beauties need a lot of water! If you find your hydrangeas start to wilt prematurely, totally submerge them in a bath of water for about 45 minutes or longer since hydrangeas also drink through their bloom.

Beware the pollen of Casablanca lilies. The pollen is quick to stain most everything they touch, but they can be easily removed. Use a wet paper towel and gently pull them off from the lily. The lily still looks beautiful, but now there's no risk of stains.

Have your phone ready to take photos of your arrangement. It's amazing how helpful this is in determining where to add flowers or what to remove to create balance.

Tulips are fragile and can droop quite easily, but did you know that can be remedied with the help of vodka? Add a splash to the water in the container and the flowers should perk up again.

Invest in an inexpensive Lazy Susan to place your container on top and go to work. It gives you easy viewing from all angles so you can see if you have any empty spots that need filling in.

Get a lush, full effect by slipping a silk orchid in with the real flowers.

Alum is a pickling spice found at the grocery store or online and can be used for persnickety hydrangeas that wilt expectantly. Dip the freshly cut tip of the hydrangea stem in the alum before adding to your arrangement for a longer-lasting bloom.

Creating a garden arrangement in your own backyard? Cut your flowers first thing in the morning, which is when they are most hydrated, and immediately submerge the stems in a water-filled vase to make them last longer.

If your non-glass vase is too deep for your flowers to rest on the rim, stuff some cellophane (such as the cellophane that the flowers came wrapped in) into the bottom to provide support and height.

For dining table flower designs, put your elbows on the table and make the height of your centerpiece no higher than the height of your wrist. Now people can still have conversations while enjoying the flowers.

PICKING
flower combinations

When I buy flowers, I am usually in a rush. Helping me narrow down my choices prevents me from overbuying and makes the process go much faster. Here are some helpful tips I use to create flower bouquets.

ONE COLOR You can never go wrong with a single color of the same flower. True perfection is a massive bouquet of pink peonies. Actually, a massive bouquet of anything is beautiful.

CONNECTED COLORS For a bouquet with more variety, I take my cues from Mother Nature. Pick a flower that has two colors in it, for example, a yellow Gerbera daisy with a pink center. For the second flower in the bouquet, look for a bloom that matches the daisy's pink center, like a pink hydrangea. If you want a third flower in your bouquet, add another pink OR yellow option, like a pink or yellow rose. You know the flowers will work in the bouquet because the colors naturally connect to each other.

MIXED COLORS If the flower you like doesn't have an obvious secondary color, then use the color wheel diagram below to help create your arrangement. (Hint: take a picture of the diagram on your smartphone so that you will have it handy when you shop.) For a beautiful contrasting bouquet, pick colors from the opposite sides of the diagram. For a bouquet with more complementary colors, choose colors that are next to each other on the diagram. For an ombre effect, start with a favorite flower, then look for flowers of the same color in graduating hues to form your bouquet. I like to keep bright colors with other bright colors, pastel colors with other pastels, but feel free to experiment with what suits your own personal taste. Remember, there are no hard and fast rules—it's all about what speaks to you personally. The page at the right has some of my favorite color combinations.

The classic color wheel.

ADDING NEUTRALS Green and white flowers act as neutrals and can be added to any flower mix. White flowers provide crisp contrast to other colorful flowers while green flowers blend softly between the blooms. Both neutrals can enhance any arrangement. White flowers are easy to find; green flowers are less so, but green hydrangeas are my go-to bloom.

ADDING GREENS Though beginners usually overlook adding greens to their arrangement, flowers pop with more vibrancy and drama when they're contrasted with green foliage. You could buy greens anywhere, or save some money and snip foliage from the backyard. Some of my favorite greens are silver eucalyptus leaves (a bonus if they have berries), lemon leaves, and of course the leaves that are already on the flowers you're using. For my purposes, I usually consider white and green hydrangeas as both greens and part of the flower selection.

ADDING TEXTURE Often forgotten, especially by beginners, textural elements add visual interest and dimension to flower arrangements. Textures act like neutrals, but they have a frilly, nubby, or bumpy appearance that contrasts to the smoothness of flower petals and leaves that is appealing. Some of my favorite types of texture are Queen Anne's Lace, Hypericum berries, rosemary, eucalyptus berries, and even flower buds that haven't bloomed yet.

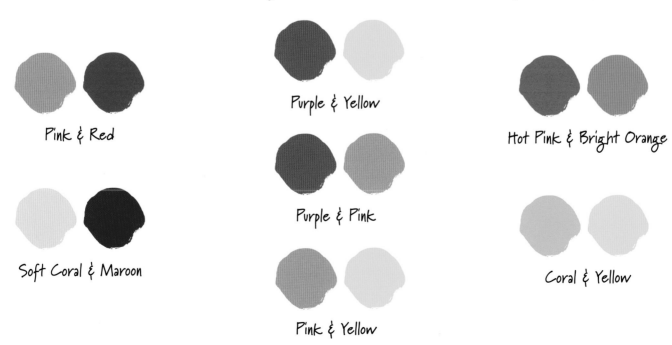

A few of my favorite color combinations
(remember to add neutrals, greens, and texture to complete the arrangement)

Pink & Red

Purple & Yellow

Hot Pink & Bright Orange

Soft Coral & Maroon

Purple & Pink

Coral & Yellow

Pink & Yellow

PREPARING
your flowers

Once you have your flowers and greens in hand, it's time to get them ready. Tempting as it may be to start arranging them in your vase, preparing the flowers and greens first helps keep your arrangement beautiful and clean, and last longer. Remember that any lower leaves on the stems that touch the water in your vase will introduce bacteria and give flowers a shorter life span.

First, use scissors to cut the stems of your flowers and greens to a manageable length. I leave on any plastic wrapping or rubber bands that may come with the flowers as I cut to ensure the stems will be trimmed to the same length. I can easily remove the plastic or rubber bands afterward.

Next, remove all the lower leaves (and snip away any thorns, to avoid prickles) on the stem that are not part of the arrangement. This means all the leaves or foliage that could be submerged in the water once you put the stems in the vase. You can cut the leaves off with scissors or use your hands for soft-stemmed flowers.

Then look at the flowers and check the petals for any unsightly or damaged ones. If they are safe to remove, gently wiggle off these petals. For roses, remove the guard petals, which are the brown, wilted outer petals of the flower, right above the sepal.

Finally, recut the stems on a diagonal so there is more surface area for them to take in water. Again, this lengthens your flowers' life span. Now you're ready to begin designing.

sepals

guard petals

EASY TECHNIQUES

MY FLOWER
equation

This is my go-to formula when creating floral arrangements. If you learn nothing else from this book, remember this: flowers + greens + texture + smile (an unexpected element). With these four basic elements, you'll never go wrong. For this arrangement, my flowers are yellow ranunculus, purple lisianthus, and white anemones; the greens are eucalyptus and hydrangea leaves; the texture is eucalyptus berries; and the smile is the fresh mint, which delivers a lovely fresh aroma. I arranged everything in milk bottles for a bright dining table centerpiece.

1 Fill each bottle three-quarters full of water. Place the eucalyptus branches, which already have berries for texture, into a few of the bottles. Make sure there are no leaves in the water.

2 Trim the hydrangea stems so that the flowers rest on the rim of the bottles when inserted, and remove any leaves that may touch the water. Place the hydrangeas around evenly in the bottles.

Cut stems in varying heights to create dimension and interest.

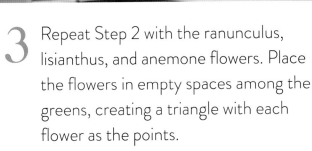

3 Repeat Step 2 with the ranunculus, lisianthus, and anemone flowers. Place the flowers in empty spaces among the greens, creating a triangle with each flower as the points.

4 Stand back to look at your arrangement for balance. Fill in any empty spaces with other flowers and greens—I used fresh mint—and adjust where color or texture needs to be added.

BUNDLING

This technique can be accomplished with two or more groups of flowers, with the option of using rubber bands to create tighter bundles. I put them in a short vase, perfect for a desktop, coffee table, or bedroom. Here, half my flowers are vibrant pink tulips with dark centers, and the other half are stark-white ranunculus for a fresh contrast. But have fun exploring other fun colors, like hot pink and bright orange, or yellow and lime green. For a grouping of three flowers, try pastel pink, lavender, and white. Or you could use three different flowers in similar shades for an ombre look.

EXPERIMENT with connecting colors when you bundle. Playing off the colors in their centers, these tulips could also be combined with light pink or yellow flowers. (See pages 24–25 for picking color combinations.)

Save time by cutting soft-stemmed flowers with the store's cellophane wrapping still on. They're already bundled for you!

1 Prepare the tulips and ranunculus flowers by cutting the stems to the same length so that the flowers will rest on top of the rim of the container. The blooms should hang over for a full look.

2 Remove any lower leaves on the stems of the flowers. Rinse off dirt and remove any loose leaves from the cut tulips. Tulips hide a lot of dirt in their leaves, so be mindful as you clean.

3 Gather the cleaned tulips together gently into a bundle and place them directly in the container. For tighter bundles, you can wrap a clear hair band around the stems first.

4 Repeat Step 3 with the ranunculus, keeping the tulips and ranunculus in separate bundles. Continue bundling and inserting until the container is full, then look at your arrangement and adjust it as necessary for balance.

REPETITION

The simple technique of repetition is a foolproof way to ensure a stunning splash of color. There are so many ways to play with this technique, but here I chose lavender sweet peas to display with delicate glass vases. You could parade them down a dining table or cluster them on a beautiful tray, like this silver one I used for contrast to add a stylish element to a bathroom. You could also use vases of the same size, but with varying heights you get a waterfall of color!

The amazing fragrance of sweet peas
adds a lovely touch to the room.

1 Fill the vases half full of water. The different heights of the vases play with the waterlines, which is part of the design; if you're using vases of the same height, you will have the same waterlines.

2 Gather a few stems of the sweet peas and hold them beside one of the vases. Visualize how long the stems should be so that the flowers will rest on the rim, and then trim.

3 Place the cut stems into the vase. Just a few stems in the vase will make the sweet peas look lush and full, but feel free to add one or two more stems if necessary.

4 Repeat Steps 2 and 3 to fill the rest of the vases. The heights of the vases will determine the heights of the flowers. Place the vases beside each other on a surface for a full display of color.

ORGANIC
roots

The look of seeing the roots of a flower inside a glass container is unique and beautiful. In this design, I used two potted aromatic hyacinth plants, a welcome sign to the start of spring. With root arrangements, flowering branches, or a single leaf, add large white or black river rocks at the bottom of your container for a grounding element.

EMBRACE the spring season by trying this technique with daffodils. Place them in a clear container similar to the pot the daffodils came in, but leave the dirt and roots just as they are for a natural look.

1 To remove the pot the plant comes in, hold the plant upside down securely in one hand over a trash can and gently tug the container away from the plant.

2 Remove any loose dirt from the bulb and roots with your hands, being careful not to break the root system. You can rinse the remaining dirt away under the kitchen faucet.

Too much water can create root rot, so be careful!

3 Add decorative river rocks to the bottom of the container for a natural elegance, then fill the container with just enough water for the roots to rest on.

4 Place the rooted plant inside the container—I added two plants to fill out the space. The roots should be barely covered with water when they're perched on the river rocks.

FLOATING
blooms

Floating flowers is easy, minimalist, and fast. There's also a lot of room for personalizing if you experiment with the waterline and play with the containers. Simply arrange your containers down the center of a table and add votive candles, and you look like a pro! Some of my favorite flowers for the Floating Blooms technique are roses, Gerbera daisies, sunflowers, and Casablanca lilies, but almost any flower will do. Here, I'm using yellow roses in candy dishes for an elegant look.

The sepals lend color contrast and provide stability so that the flower doesn't fall apart.

1 Remove the guard petals from the roses (the outer bottom petal that looks brown around the edge or wilted). If you're using other flowers, then make sure the petals look nice and healthy.

2 Snip off the rose heads using scissors, or snap them off with your hands. Make sure to leave on the sepals (the little green leaves under the flower bud) to keep the petals together.

3 Fill the containers with water. With clear containers like these, the waterlines are part of the design, so feel free to play with the water level. Here, I went with a low, one-inch waterline.

4 Float the roses in the water of a container, placing one flower head at a time. Place as many as needed in order to keep the flower heads facing up. Then repeat for the rest of the containers.

THE COLLAR

Adding a wreath or collar around a bouquet is a great way to add dimension and with little effort. You can create a collar using the flower's own natural foliage, different greens, or even another flower. Imagine fluffy white petals lining the top rim of a vase with tall green flowers elegantly standing in the middle—beautiful! Here, I used white parrot tulips, one of my absolute favorite flowers. I combined these showy blooms with delicate grevillea stems for the collar and secured them with a clear hair band. I chose a low white vase to keep the look simple and monochromatic.

Tulips are fragile and delicate, so be gentle when handling them.

1 Prepare the flowers by gently removing the leaves from the tulip with your hands. Tulips hide a lot of dirt in the leaves, so rinse any dirt from the stems to keep the glass container clean.

2 Gather the tulips together in one hand with a firm grip, holding the stems right under the flower heads. With your other hand, add one grevillea stem at a time all around the circumference of the tulip bundle.

3 Adjust the bouquet in your hands if necessary, making sure the grevillea leaves reach out further than the flowers to create a collar. Then cut the stems so that the tulips will rest just on the top of the container.

4 Use a large clear hair band to secure the bouquet tightly, adjusting the grevillea stems if needed to emphasize the collar. Fill the container half full of water and then place your arrangement inside.

BRICKLAYING

Bricklaying is a fast way to create a lush bouquet—and all you need is a rubber band. Because the pattern looks so regular, it looks like the bouquet took a lot of time to put together, but the arrangement is actually deceptively simple. This technique is also a shortcut for the traditional hand-tied posy! I like using one kind of flower to showcase its beauty, and picking a clear vase allows the stems to become part of the arrangement. Here, I've used two dozen soft coral roses, but firm-stemmed tulips, calla lilies, and carnations all work well. Just remember to get lots of flowers because you will need more than you think.

The key to this technique is to keep a pyramid in mind as you stack.

1 Remove the leaves from the flower stems and any guard petals, if necessary. Arrange a row of seven stems on a work surface. Then stack on top six flowers in the spaces between the flowers in the row below.

2 Continue stacking with one fewer flower in each row, until you use all the flowers you have prepared. Carefully but firmly, gather together the flowers and wrap a large clear hair band halfway up the stems.

3   With the bouquet in one hand, cut the stems straight across at a height so that the flowers will rest on the rim of the vase. Then adjust the flowers as needed to create a nice top shape.

4   Fill the vase half full of water. With the hair band still on, use your hands and gently give the stems a twist for a curved look, and then place the flowers in the container. Adjust as needed.

THREE
corners

This technique plays with geometry and balance. It's all about building your arrangement upward from the container and then filling in the open spaces. The key is to imagine an invisible triangle as you design. You can use any type of flowers, but I recommend hydrangeas as a lush foundation. Here, I play with Snow on the Mountain greens, white garden roses, Casablanca lilies, and Hypericum berries for some texture and a pop of color.

Create the triangle points out of three big hydrangeas, or three big bundles made up of small hydrangeas.

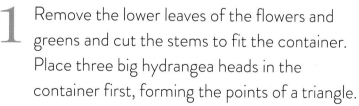

1 Remove the lower leaves of the flowers and greens and cut the stems to fit the container. Place three big hydrangea heads in the container first, forming the points of a triangle.

2 Fill in the spaces between the hydrangeas with three bunches of green leaves. Nestle in the berries with the greens, then add three bundles of white roses into the empty spaces.

3 Top off the arrangement with Casablanca lilies right in the middle of the triangle. If the lilies are open, be sure to remove the pollen (see page 21) so it doesn't stain anything.

4 Step back from the arrangement and look at it from all sides. Check to see if there are any holes that can use more ingredients or if anything needs to be moved for balance.

INSIDE
the container

This idea has so many possibilities! Once you find two clear containers that are the same height but one narrower than the other, then you're in business. Fill the space between the two glasses with any fun ingredient—jelly beans, wine corks, seashells, even decorative wrapping paper. For the flowers, I suggest using two to three similarly colored flowers with different petal textures. I sliced pink grapefruit to pair with alstroemeria, hydrangeas, tulips, and carnations. To add texture, I found Hypericum berries and white Veronica flowers.

If using small items like jelly beans, cover the inside of the container with plastic wrap to keep it clean as you fill the sides.

1 Fill the narrower container three-quarters full of water and then place it inside the wider container. Fill the space between the containers with your preferred material— I used sliced grapefruit here.

2 Remove the leaves from the flowers and cut the stems to size. Place the hydrangeas in the water in the narrower container, making sure the flowers hang over the rim of the wider container.

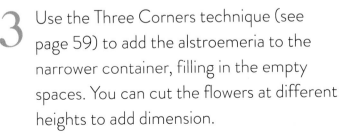

3 Use the Three Corners technique (see page 59) to add the alstroemeria to the narrower container, filling in the empty spaces. You can cut the flowers at different heights to add dimension.

4 Poke in the tulips in groups of two or three, gently threading them into the arrangement so that they hang over the other flowers. Lastly, insert the berries, carnations, and Veronicas.

OUTSIDE
the container

This technique is always a crowd pleaser, and it's easier than you think. Small cylindrical glass containers work best and can be covered with anything. All this technique requires is a rubber band and some raffia. Here, I used tulips with asparagus on the outside, but you could try rosemary, or strong stems cut from leftover roses or sunflowers. You can also wrap the vase with large hydrangea leaves, cinnamon sticks, or even candy canes!

Check the water daily because tulips like to drink!

1 Hold an asparagus stem next to the container to visualize where to cut it. Then trim the asparagus stems so that they are long enough to cover the container with a few inches of extra length.

2 Place a wide rubber band around the center of the container. I used the rubber band that the asparagus was bundled in at the market. Insert the asparagus stems, one by one, inside the rubber band.

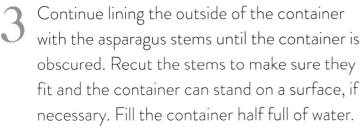

3 Continue lining the outside of the container with the asparagus stems until the container is obscured. Recut the stems to make sure they fit and the container can stand on a surface, if necessary. Fill the container half full of water.

4 Wrap some raffia around the rubber band until the rubber band is completely hidden, and tie a bow to secure. Clean and trim the tulips, then add them to the container, letting them cascade over the sides.

EVERYDAY IDEAS

BRIGHT
yellow trio

Layer shorter stems on top of the longer stems.

This idea of repetition (see page 39) plays with the bright, cheery yellows of three different kinds of flowers! Here, I chose ranunculus, calla lilies, and daffodils resting on top of philodendron leaves for a simple touch. This idea also works well with flowers sharing the same color but in varying shades.

CHOOSE three of the same vases and fill each one-third full of water. I like this oblong shape to showcase the stems as part of the design. Then prepare the flowers by removing the leaves and cleaning any dirt or debris from the stems.

DRAPE one of the flower types diagonally into a vase, cutting the stems as needed to create a cascade. Repeat with the other flower types in their own vases, being sure that the bottoms of the stems are submerged in the water.

TUCK a large leaf (I used a philodendron) into one container for the leaning flowers to rest on for a finishing touch. Repeat again for the other two containers. The leaf is optional, but it adds a little touch of elegance.

ELEGANT
pink ombré

For this look, I was inspired by one of my mentors, celebrity florist Jeff Leatham. Here I created my own spin, using the Bundling and Bricklaying techniques (see page 35 and 55) and letting one flower lead to the next flower choice. I used coral roses, pink phalaenopsis orchids, and magenta hydrangeas. This is a great piece for an entryway, console, or buffet dining.

PREPARE the flowers by removing the lower leaves and gently wiggling off the guard petals from the roses if necessary. Fill your container halfway with water.

BRICKLAY the roses and secure the stems with a rubber band. Holding the roses next to the vase, find the best height and cut the stems. Place them in the vase.

REPEAT with the orchids. After cutting the stems, place them in the vase next to the roses. Add the hydrangeas to the front of the arrangement to fill any empty gaps.

SPRING BRANCHES
with a lily collar

Blooming branches are a sign that spring is in the air. Nothing makes me happier than having a flowering branch in a glass vase with white river rocks, or this ceramic vase. Since this vase would hide any stems, it's easy to add a little something extra with a flower collar (see page 51), like these Casablanca lilies. I love the organic look of unopened buds. Here, the lilies complement the buds on the cherry blossom branches.

FIND a container large and heavy enough to hold the branches. Depending on your space and the size of your container, you may need to cut the branches.

FILL the container three-quarters full of water and place the branches inside still secured with the rubber band they came with. If the branches don't look good like this, then cut the rubber band and arrange in the container.

REMOVE all the lower leaves of the lilies. Hold them next to the container and cut the stems so that the stems reach the water. Place the stems evenly around the branches so that they hang slightly over the edge of the container.

STAND back and make sure everything looks balanced, and adjust where necessary. Be sure to check the water level daily to keep the lily stems in water.

A BOWL
of tulips

I love the unique and effortless look of resting flowers inside a bowl. Using a fish bowl works well for this design with flowers that bend easily, like calla lilies or tulips, which I used here. Display this arrangement on a coffee table, or use three bowls for a splash of color down your dining table.

Remember to wipe down the fish bowl for a clean finish.

FILL a bowl with enough water so that all the stems can be submerged. Remove the majority of lower leaves from the tulips, keeping just one or two leaves on for some color. Rinse the stems of loose dirt. Hold the stems together and cut them evenly.

WITH your thumb and index finger starting from the top of the stem, gently press to create a natural bend to the stem. Be gentle, but don't be afraid; it's okay if some stems break. Many of the stems may be already curved anyway, which is convenient.

PLACE a tulip inside the bowl at an angle and wrap it so that the stem is in the water but the flower and leaves are not completely submerged. Turn your bowl slightly and repeat. Continue so the flowers are in the same direction until you complete the circle.

DOUBLE
duty

There are many decorative trash cans, so why not repurpose them? For this arrangement I found a lovely burnished gold waste can that serves as a stylish container. Here, I used eucalyptus branches and the variegated leaves of a dragon tree (dracaena)—I love these types of leaves because the eye goes to the color white first, making the arrangement more dramatic—coupled with delicate pink roses.

FILL the container half full of water. Clean off the lower leaves from the flowers and greens, and gently take off the guard petals for the roses if necessary.

CREATE the structure by starting with the eucalyptus, placing bundled eucalyptus branches inside the vase. Then add the variegated leaves. I like adding them at different heights for dimension.

TRIM each rose stem individually, then insert the flowers one at a time in the vase to create an angle or line that moves the eye upward.

SIMPLE
repetition

Sometimes the vase makes the flowers—or is it the other way around? Either way, this arrangement is a showstopper! I adore these vases because the glass is like artwork and needs very few ingredients to create something fast yet beautiful. I used the Repetition technique (see page 39) for these white lisianthus, a favorite because they bloom all year round, are inexpensive and long lasting, come in many colors, and have a simple elegance.

Unopened buds add a wonderful element of texture.

REMOVE the lower leaves from the bottom of the stems that could be seen in the glass. Cut the stems so that the top leaves are just above the rim of the glass.

FILL the vases one-third full of water. With clear glasses, the waterline is part of the design, so be sure to keep the waterline the same for each vase.

ADD one to two stems of flowers in each glass for a flowy look. Stand back and adjust to make sure the glasses are evenly balanced. Display in a line for the maximum effect.

GARDEN
style

The popular trend of an unstructured garden look feels fresh, simple, and organic. This style often pairs with a pedestal vase, which always raises the beauty bar. I like to start the design with greens, add smaller flowers, and then pop in some focal flowers. This monochromatic look of yellow tulips, garden roses, and showstopping peonies plus eucalyptus leaves and bupleurum is sure to please!

FILL the vase half full of water. Place tape strips across the mouth of the vase in a half-inch grid pattern, then tape the outer rim of the container to secure the grid.

REMOVE the lower leaves of the greens and flowers. Insert the eucalyptus branches and bupleurum in the grid first, placing them to stand taller on one side of the vase than the other to create movement.

ADD the roses, with a few of them cut shorter so that they are close to the base to hide the tape grid. Vary the height so that the flowers have some movement as well.

POP in the tulips and peonies last, keeping at least one flower at eye level for a nice focal point. Then stand back and fill in empty spaces with more greens or flowers.

DAISY
in a bottle

It is so simple to reuse any pretty water bottle for a quick breakfast arrangement. Here, I used three bottles to echo the Repetition technique (see page 39) and a serving platter as my tray. Add some flowy ornamental oregano and a single-color Gerbera daisy, and you are good to go.

CHOOSE three similar bottles and fill each half full with water. Make sure they all have the same waterlines, which is part of the design.

REMOVE the lower leaves from the greens that could end up inside the bottle. If some of the daisy petals look damaged, remove those petals and save just the flower centers for an extra design element.

ADD a couple sprigs of greens in each bottle. Finish them off with a daisy in each bottle, and display the bottles together in a line. Put any flower centers on the platter.

SINGLE STEM
r o s e s

Simple flowers plus easy design equal big impact! A vase with a large opening would normally require many flowers to fill, but just a few angled single-stemmed roses and some river rocks do the trick for a fast and easy arrangement. Here, I used three hurricane candle holders, two tall and one small, for a high-low effect.

Recreate this with any firm-stemmed flowers, such as sunflowers, gladiolas, and calla lilies.

FILL each of the vases with a few inches of water, creating a low waterline for a clean look. Gently add a single layer of black river rocks for another element of depth.

REMOVE the lower leaves on the rose stems, leaving on a couple leaves closer to the bloom to provide contrasting color and texture. Gently remove any guard petals from the roses.

INSERT four to five stems into each vase in a crisscross pattern, leaning them against the vase. The stems could be different heights, but the flowers should always be above the rim of the vases.

ROSEMARY
bouquet

Using rosemary for the Outside the Container technique (see page 67) in this design makes it feel like the flowers are growing right out of the garden! For this, you need a small container with enough rosemary to cover the outside, a rubber band, some raffia, and some flowers. I used light coral roses, yellow and white spray roses, sunflower centers, and love-in-a-mist.

WRAP a thick rubber band around the container. Remove the lower leaves from the greens and flowers and cut the stems: varying heights for the greens, shorter stems for small flowers, and longer stems for large flowers.

INSERT rosemary stems inside the rubber band until the container is obscured. Hold the container in one hand and use scissors to cut the bottoms of the rosemary stems evenly with the container's base.

COVER the rubber band with raffia, looping it several times around, and then tie it off with a bow. I like to cut the loops of the bow for a more casual look. Then fill the container half full of water.

ADD the greens to the container first, angling them upward and outward towards the sides. Then tuck in clusters of small flowers in the empty spaces. Lastly, add the large flowers as focal points.

FLOATING
apples & roses

For this idea, look around your home for your favorite wide, low containers. I love this silver bowl for its reflective quality, especially when I use it for floating flowers (see page 47). Floating arrangements require little to no prep because you just use the flower head, making everything fast and easy. You can add fruit, floating votives, or just use all flowers. Here, I used 12 rose heads and 7 crabapples.

You can also use other fruit, like limes and lemons.

FILL the container with enough water so that the apples and flowers float. I filled my container about three-quarters full.

TAKE off the guard petals from the roses. Snap off the rose heads with your hands, or use scissors to cut them off.

PLACE the apples in the water alternating with the rose heads. I like to see more flowers than fruit in the bowl.

CLASSIC
calla lilies

Flower possibilities can be overwhelming, so when in doubt, go with a single-flower bouquet. This simple arrangement plays with the Bricklaying technique (see page 55), using a tall clear vase, white river rocks, and beautiful calla lilies. This is a gorgeous and welcoming addition to any entryway.

For soft-stemmed flowers, layer with equal stems facing left and right as you build the pyramid.

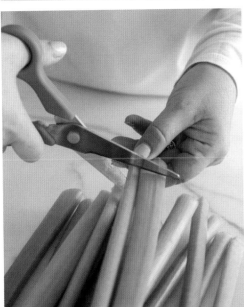

FILL your vase with water and then gently slide the river rocks into the water. I like a low clean waterline for a contemporary look. Less water also limits bacteria growth.

RINSE the stems of the calla lilies to remove any dirt. Begin bricklaying the flowers by placing six stems for a bottom layer, five stems for the next layer, etc. until you create a pyramid.

GATHER the stems together in a firm grip and then cut the stems to your preferred height by holding the flowers next to the vase and looking for a height that appeals to you.

PLACE the flowers in the vase and give the stems a gentle twist so they look slightly turned. You'll want to keep an eye on your flowers to see when they may need another drink.

THREE BY THREE
sunflowers

This is the cheeriest—and easiest—display that anyone can put together in a snap. All you need are some gorgeous flowers, like these showy sunflowers, and the same type of candy dishes, candle votives, or any small containers. Using the Floating Blooms technique (see page 47), float individual flower heads in the dishes, then display them down a table with tea candles, or place them together in a tray for a mass of color.

CUT the flowers from their stems with scissors. The stems should be long enough to reach the water but short enough for the flower head to be able to rest in the votive.

FILL each small container with enough water so that the flower will float. If you're using clear containers, make sure the waterlines are all the same.

PUSH the flower petals away from the center to gently fluff them, and remove any damaged petals. Place a flower in each container and display together.

STANDING
roses

This creative idea repurposes parts of the flowers you typically throw away. Here is a design using a tape grid to keep flowers in place while reusing the flower stems in a unique way. I like placing this arrangement on a coffee table, side table, or down a dining table where you can easily see the grid pattern that might otherwise get hidden. I prefer using waterproof tape, but regular tape works fine.

FILL a low, square container about one or two inches with water, creating a low waterline. Remove the leaves and guard petals from the roses if necessary, but leave on the sepal.

TAPE strips across the top of the container to create a grid. It's okay if a small amount of tape is over the edge. Cut the rose stems so that they reach the water while the flower rests on top of the grid.

INSERT the rose stems into the spaces of the grid until the container is filled up and the roses are touching comfortably on top. The rose petals will hide the tape.

MEASURE leftover stems and cut them to place on top of the tape strips of the grid. Now the flowers are seemingly resting on a stem grid rather on a tape grid.

MIXING
plants & flowers

Here's a fun hack that takes just a few minutes to put together but makes you look like a master gardener. This flower arrangement is actually an array of potted plants in a long trough—all you need to do is keep the plants watered and switch out the blooms as needed. My plants were Philodendron Birkin, Angel plant, orchids, and hydrangeas, and I paired them with cut tulips and daffodils.

Bring your container to the store to make sure you buy the right number and size of potted plants.

NESTLE flowers and greens in a trough or another long, narrow container. Feel free to play with different heights, volumes, and types of flowers and greens.

CREATE some room in the front so that you can add a plastic cup filled with water. Be sure it is near the front and hidden by the surrounding potted plants.

INSERT freshly cut flowers into the cup, like these pink tulips. As the flowers wilt, replace them with a new swatch of bright blooms, such as daffodils.

FLORAL
mojitos

I use a tray for display, but you can also line the vases down a table.

I try to look around my home for vases by seeing what I can repurpose. It saves on vase storage, it saves on time, and it makes me feel quite clever. I bought this Mojito set for entertaining, but I love it so much that it mostly gets used for flowers. Here, I used lisianthus, lilies, and dahlias.

REMOVE any lower leaves from the flowers and cut the stems to size. Fill each glass half full of water. How many flowers you put in each glass depends on the size of your blooms.

CREATE three equal-sized masses of colors, from lightest to darkest color. I added the lisianthus to three glasses, the lilies to two glasses because they are so big, and the dahlias to the remaining three glasses.

TAKE a step back and view the arrangement from all angles to check the balance of the flowers and colors. Add or remove flowers where necessary, and vary the heights of flowers for depth.

GREEN APPLES
with flowers

This low design is perfect for a summer brunch and it won't impose on conversation. All you need are some flowers and apples, and clear hair bands to create your bundles (see page 35). I love the happy color combination of bright green and yellow. Here, I used yellow baby carnations, white stock, and Solidago to pair with the green apples.

Bring your container to the store to buy apples. You'll need more than you think.

PREPARE the flowers and greens by removing any lower leaves on the stems. Clean the apples so that they are ready for display. Then fill the container half full of water.

CREATE a few mixed bouquets of flowers and greens by pulling from each ingredient, making enough bundles to fill the container. Secure each bundle with a rubber band.

HOLD one bundle and cut the stems so that the flowers hang over the rim of the container when placed in the container. Repeat for the remaining bundles with most of the remaining ingredients.

SET the arrangement on the table and gently stand green apples upright around the container. Fill in any obvious holes with more flowers or greens.

TULIPS
on ice

Your friends will definitely do a double-take of this showstopper! The secret here is good old-fashioned gift-wrapping cellophane to create a cool effect. In this arrangement I used two cylinders, one large and one medium, with cymbidium orchids because they work well in water, as do tulips, roses, and succulents. I also added three dozen purple tulips to top off my design.

Less cellophane makes for a lighter crushed ice look.

FILL the container one-third full of water. Use scissors to cut several pieces of cellophane. Then prepare the flowers by cutting the orchids from their stems and cleaning the tulip stems of any dirt.

CRUMPLE and push a piece of cellophane gently into the water. Carefully place an orchid inside, pressing it against the inside of the glass vase. Repeat until the container is almost full.

FINISH by cutting the tulip stems, keeping longer stems on the bottom and shorter stems on top. Add the tulips to the top of the vase at an angle, making sure the bottom of the stems are in the water.

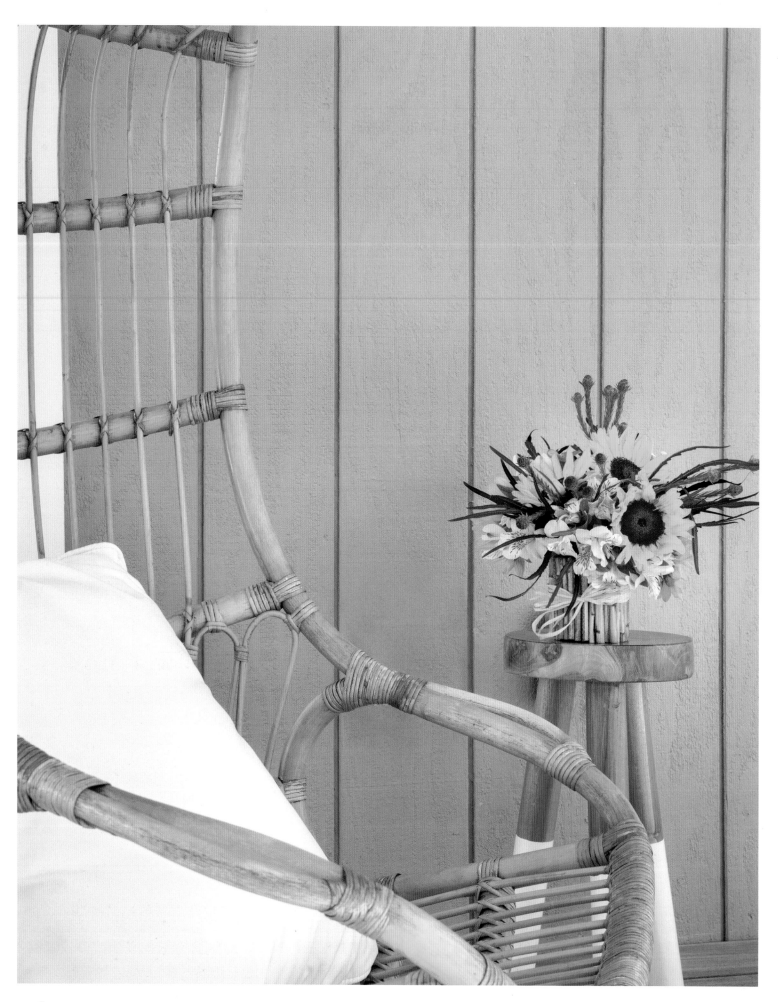

SUNFLOWERS
inside & out

When it comes to ideas to use for Outside the Container (see page 67), reusing stems from your flower is a convenient way to add uniqueness to your design. All that's required is a thick rubber band and some raffia. Aside from sunflowers, I also used alstroemeria here with black mondo grass and ornamental pine for texture.

WRAP a rubber band around the container. Sunflower stems are long, so cut enough stem pieces to go around and cover the length of the container, being sure to leave enough stem for the sunflower in the arrangement.

INSERT the cut sunflower stems under the rubber band so that they completely cover the container. Then cut the bottoms and tops off the stem evenly. Fill the container half full of water, and prepare the other flowers and greens.

PLACE the alstroemeria inside the container at an angle so that the flower rests on the rim, and the stems create an internal grid. Next, add the sunflowers, mondo grass, and pine in the grid to fill in the holes.

HIDE the rubber band by wrapping the raffia around it and tying it in a bow. Snip the loops for a less fussy look. Take a step back and fill in any empty spaces you may see.

SHOWER
greens

Love is in the details! If friends or family are visiting and find flowers in the shower, it would certainly put smiles on their faces. Having leftover flowers from another arrangement, I created some simple swag using spa-like fragrances of eucalyptus, rosemary, and white lisianthus. It's easy—all you need are a rubber band and some raffia.

You can also add other aromatics, like lavender or lemongrass.

PLACE eucalyptus branches on the table and cut them to your desired length. The eucalyptus determines how long the bundle will be. Add a few shorter stems of rosemary and a couple of flowers.

GATHER the bottom of the stems together, making sure they're even, and hold the bundle tightly. Secure with a rubber band about three inches up from the bottom of the stems.

WRAP a long piece of raffia around the rubber band and double-knot it to secure. Then take the two loose ends and tie it around the showerhead to hang. Cut off the excess raffia.

BEDSIDE
hydrangeas

You can never go wrong with one kind of flower for a splash of color—it's lush, elegant, and always beautiful! Bedside posies are a personal favorite. The flowers are like my little oasis of beauty to enjoy before falling asleep and as I wake up. All you need is a small vase for a single-stemmed flower or some hydrangeas (which I used here), and a clear hair band.

FILL your vase half full of water. Gather a few stems of hydrangeas to create a nice posy for the size of your vase, then remove the lower leaves from the stems, keeping the upper leaves attached.

HOLD the flower stems together next to the vase to estimate the length. Use scissors to cut the stems so that the flowers and the upper leaves will rest on the rim of the vase.

SECURE the stems with a rubber band and push it up toward the upper leaves so that the blooms are snug. This is particularly helpful to hold the flowers in place when using a short vase.

ADJUST the upper leaves under the flowers, gently moving them to create a natural collar while making sure they're still attached to the stems. Then place the posy inside the vase.

SUNRISE
tulips

When I saw them at the farmer's market, I could not resist buying several bouquets of these fiery orange parrot tulips with their dark centers. In this case, I repurposed my tall glass hurricane candle holder and used the Bricklaying technique (see page 55). Don't be surprised if some tulips close a little at night; they'll reopen in the morning. Tulips also continue to grow after they have been cut.

Tulips drink a lot of water, so a glass container makes it easy to keep an eye on the water level.

FILL the container one-quarter full of water. Prepare the flowers by removing most of the lower leaves off the flower stems and gently rinsing the stems of any loose dirt that may be hidden.

GATHER one bunch of tulips as they came from the store, or use the Bricklaying technique. Cut the stems so that they will still reach the water when they are leaning, and place them in the container.

REPEAT with another bunch of tulips until the container is full, leaning each bunch over another. With all the tulips leaning, the arrangement creates a beautiful cascading effect of flowers.

RESOURCES

I visit the places below to get all the tools and materials I need to create flower designs, but more generally you can find everything you need at the following stores:

GROCERY STORES Most have flower sections, and you could also buy baby bottle brushes, alum, etc. there and repurpose the rubber bands if you buy produce.

KITCHENWARE STORES These would have great options for containers (glasses, vases, candy dishes, bottles) and probably scissors.

HOME AND FURNITURE STORES Containers (candle holders, baskets, lanterns) can offer their intended use as well as for holding flowers. You might also find river rocks.

CRAFT STORES These have scissors, raffia ribbon, tape, and rubber bands.

ONLINE STORES When you can't get it local, go online!

ANTHROPOLOGIE
(anthropologie.com)
I love Anthropologie's homeware section where many items can be repurposed for vases. Not every store has an amazing garden shop, but the ones that do, watch out. It also has great artificial flowers.

AMAZON (amazon.com)
For a one-stop shop, you can find almost all your tools and supplies on Amazon. It's a convenient place to get alum powder, clear elastic hair ties, non-skid Lazy Susans, and kitchen scissors. I also love getting variegated tea leaf–patterned waterproof ribbons to line the inside of my glass vases.

CRATE & BARREL (crateandbarrel.com)
CB2 (cb2.com)
Both stores have wonderful vases. I love browsing the kitchen glassware section to see what can be repurposed for vases: nut bowls, olive dishes, and other small bowls, which are perfect for candles and floating flowers. They also have a lovely garden section.

HOMEGOODS (homegoods.com)
This is a treasure hunter's dream. Great price point and awesome selection of lanterns, containers, vases, and much more. Set aside time to browse it all.

HUDSON GRACE
(hudsongracesf.com)
Beautiful high-end home store with amazing containers and some great prices. Buyers beware; you may go home with much more than anticipated.

MICHAELS (michaels.com)
You can get most everything here—raffia, rubber bands, ribbon, river rocks, vases, alum, scissors, baskets, and much more, all at affordable prices.

POTTERY BARN (potterybarn.com)
WEST ELM (westelm.com)
Both stores have great vases. I get my tall cylinder vases here which I use for buffets, console, and entryway arrangements. Also check out the candle section for container options. Both stores also have great artificial flowers.

SAFEWAY (safeway.com)
Their prices and variety are very good. You can also pick up supplies such as barbeque skewers for placing fruit in arrangements and a baby bottle brush for cleaning narrow vases. You can also find the alum powder, which helps hydrangeas last longer. Make sure to save the rubber bands from your vegetables for Outside the Container designs.

TRADER JOE'S (traderjoes.com)
They have the best prices. If you have an event, give the flower buyer a heads up and they can order in advance and in bulk.

WHOLE FOODS MARKET
(wholefoodsmarket.com)
Whole Foods always has dependable and healthy flowers. Don't overlook the potted flowers which can be cut down, provide alternative colors, and may be less expensive.

WORLD MARKET
(worldmarket.com)
World Market has great price points for one-stop shopping. Browse their vases section where you can also find white and black river rocks. Check out the candle department for lanterns to use as vases. They also have great baskets. I love the kitchenware section for water or beer glasses to repurpose for individual flowers down the table. You can also find raffia in their gift wrap section.

ACKNOWLEDGMENTS

I am extremely grateful to all those who have supported me along this journey!

To Jennifer Newens at West Margin Press for providing the opportunity to create a book on simple flower designs for everyday living, and for guiding me through the publishing process. It is with your grace, kindness, and talented team members, Rachel Metzger, Olivia Ngai, Angela Zbornik, and Michelle Montano, that has made this journey so rewarding.

To Victoria Zenoff, career coach and life strategist, for guiding me, believing in my passion, and for putting this book puzzle together.

To my photographer, Vivian Johnson. Fate brought us together for a reason. I am endlessly grateful for your hard work and dedication to excellence. You are an amazing talent. This book would not be the same without you!

To my dear friends, Pam and Jim Robertson. Thank you for generously offering your beautiful Belvedere home for my photo shoots. Your spa-like interiors showcased my designs with perfect lighting and inspirational views of the San Francisco Bay. Pam, I am lovingly indebted to you for your never-ending help and friendship, your endless energy, and your talented eye for styling. I could not have done this without you.

To my dear friends, Maggie and Stephen Oetgen. Thank you for offering your incredible Napa home for a most inspiring photoshoot. A home with such impeccable décor, surrounded by lush vineyards—I never want to leave!

To Brian Woolery at the Ink House Inn in Napa, for generously offering our crew a last-minute photoshoot opportunity!

To Darice O'Neill for your loving friendship and for starting me on my flower journey over 20 years ago. By sponsoring me for the Ross Valley Women's League, we volunteered for almost a decade for Adopt-A-Family's sold-out annual auction. Here, we were given unlimited encouragement to express over-the-top, theme-filled flower designs. A platform that parlayed into the flower enthusiast I am today!

To Randy, my devoted husband of 30 years. Thank you for such a beautiful life. Thank you for supporting me in everything I do. You are my rock. I love you so much!

To my three amazing children, Kendell, Courtney, and Taylor. Thank you for encouraging me to dream big and for always cheering me on along the way. Your creative comments have made all the difference in the making of this book.

To Pierre, my first son-in-law, you set a high bar! To my godson Cole, you fill my heart like no other! To my other daughters, Makena, Alex, Becky, and Justine, forever family!

To Pam, my loving sister and biggest influencer. Your talent is endless and a true inspiration, your generosity always without limits. What would I do without you? To Brad, my loving brother-in-law, thank you for modeling inspiring strength, courage and a generous heart!

To the rest of my amazing inner circle of friends from California to Florida. It is said that family is anyone you can't imagine living your life without. Well, I am blessed to have a very large family! To our 30 years of friendship, Amy, Marci, and Debi, your love carries me through life. To Kristen, Lynn, Katie, Cyndi, Deanne, and all my loving friends, new and old, who have supported me with years of encouragement, continued oohs and ahhs, phone calls for flower advice, and inspiring flower photos from your travels. You fuel my creativity and fill my heart. To all of you, husbands and children included, I am forever grateful. It is because of you that I live a truly blessed life!

In loving memory of my mother, who always encouraged me to go after life by creating a foundation of love where nothing was impossible.

INDEX